DEPARTMENT OF DOCKS AND FERRIES
CITY OF NEW YORK
PIER "A," NORTH RIVER

REPORT

ON

Transportation Conditions

AT THE

Port of New York

WITH

ESPECIAL REFERENCE TO A JOINT RAILROAD TERMINAL IN MANHATTAN ON THE NORTH RIVER ABOVE 25TH STREET.

SUBMITTED BY

CALVIN TOMKINS,

Commissioner of Docks.

JULY, 1910

1837
ARTES
SCIENTIA
VERITAS
LIBRARY OF THE
UNIVERSITY OF MICHIGAN
TUEBOR
CIRCUMSPICE

No. 2A.

DEPARTMENT OF DOCKS AND FERRIES
CITY OF NEW YORK
PIER "A," NORTH RIVER

REPORT

ON

Transportation Conditions

AT THE

Port of New York

WITH

ESPECIAL REFERENCE TO A JOINT RAILROAD TERMINAL IN MANHATTAN ON THE NORTH RIVER ABOVE 25TH STREET.

SUBMITTED BY

CALVIN TOMKINS,

Commissioner of Docks.

JULY, 1910

Hon. William J. Gaynor,

Mayor, and Chairman of the Board of Estimate and Apportionment.

Sir—Pursuant to your instructions, I have for some months past been seeking information and co-operation from transportation interests and commercial bodies relative to a terminal policy for the Port of New York. I had expected to report on this matter in September of this year, but the discussion of the west side surface track problem has immediately necessitated a preliminary and partial consideration of the subject, which I trust may be received and considered as a tentative basis for criticism and negotiation, not as a conclusive study.

I am pleased to report that the transportation companies with whom I have consulted, including all of the railroads and most of the water carriers, have shown a lively interest in the suggestion for modern connected terminals for joint use, and have indicated their willingness to negotiate and co-operate with the City to relieve the increasing congestion with its attendant expense and inconvenience.

The west side problem is the most important single factor connected with the general port policy of the City. There are, however, many other matters to be coincidently considered, and the more important of these are so involved with this that they must necessarily receive some present attention. I shall expect to return to a more comprehensive discussion of the entire subject in the Annual Report of the Dock Department for the current year.

Rate Discrimination Against New York.

The unrivaled harbor of New York and its situation at the end of the only level route from the Mississippi valley to the Seaboard constitutes the basic reason for its supremacy as the metropolitan seaport of North America. Other seaport cities and the railroads leading to them have for 28 years sought to neutralize these natural advantages by imposing discriminating freight rates on commerce to and from this City. Railroads terminating here have entered into this conspiracy, and until recently public opinion in the State and City has acquiesced, has permitted the Erie Canal to fall into decay and has ignored its responsibility of providing modern water and rail terminals within the Port.

At last the leading commercial bodies of this City have combined to terminate this unjust and harmful discrimination. The railroads and the Interstate Commerce Commission will be appealed to.* Public opinion in the West, which section is also suffering from its inability to fully avail of the cheapest route to the seaboard, will be asked for co-operation.

Finally, the Erie Canal improvements now under way, together with the terminals which the State will doubtless provide at New York and Buffalo, in addition to the terminal facilities which the City should provide for the vast commerce seeking its harbor, may be depended upon ultimately to compel the railroads to permit the City to fully utilize the advantages of its natural opportunities.

New York must promptly modernize its waterfront terminals for joint rail and marine use, instead of leasing its docks for long terms to individual corporations, no one of which is providing or can provide them with suitable equipment. The docks have been unintelligently operated as separate units for the reception, delivery and interchange of freight, without regard to the joint need of carriers or the general commercial needs of the City.

The development of manufactures within and in the immediate vicinity of a port is of far greater importance, locally, than the passage of commodities through it in transit. It is becoming imperative that modern factories receiving and sending out heavy or bulky commodities should be provided with railroad switches over which connection can be made to all roads serving the community. Cheap terminal facilities are as necessary as favorable freight rates for successful industrial competition.

Importance of Terminals.

President James J. Hill, in a recent interview, states that "the pressure upon existing terminal facilities is a future menace and a present handicap. For months past it has been impossible to get freight shipments delivered promptly, if these have to be transferred at any of the principal terminal points. * * * The

---

* The City should be represented by the Corporation Counsel in these proceedings.

flood of business which rose to such dangerous heights in 1907 is piling up again, with the additions made by national growth since then. * * * The future will add to these difficulties as well as to the losses which they involve. * * * The only probable relief upon terminals, where the greatest difficulty exists, is the decline of our export trade. * * * The problem of terminals is the greatest in the country; it is the problem of transportation agencies, of financiers, and of the communities directly affected, and of all the interests that depend upon cheap and speedy carriage for the commodities which they buy and sell."

Increasing land values, and the difficulties attending changes in established City plans, have made it well-nigh impossible for the railroads to materially extend their terminals in large cities without municipal co-operation.

Railroad experience shows that local terminal charges at cities situated relatively to each other as are Philadelphia and New York, aggregate many times the actual cost of transportation between such cities. Great economies have been effected by railroads in alignment, strengthening of bridges, use of heavier rails, larger, heavier and more powerful locomotives and cars, but terminal charges have not been correspondingly reduced.

The completion of the Erie Canal improvements, and later on the opening of the Panama Canal, and the prospective provision for an intercoastal canal from Baltimore, through New York harbor to Fall River or Boston, will inevitably swell the volume of commerce flowing through New York, and means for handling this increase should now be anticipated.

Local Difficulties to Overcome.

Many cities are so situated that pressure on terminals can be relieved by circular freight roads connecting separate freight stations and providing transfer points for less than carload freight. This solution is not at present possible in New York, where the problem is greatly complicated by water separation of the Port into four parts, namely, the New Jersey Section, the Manhattan and Bronx Section, the Long Island Section and the Staten Island Section. Freight communication between these four parts is now provided for by car floats and lighters; and not until the community is rich enough to afford freight as well as passenger tunnels under the rivers will the consequences of this separation be entirely overcome. The municipal problem at present is to provide for waterfront terminals, where transfers can be quickly and cheaply made between cars and vessels, and between different railroad lines.

The natural features of the Port are too big for full immediate utilization. They constitute, in fact, an embarrassment of opportunities. The present policy should be one which shall seek to make the best use by car floats of terminals so located that they can subsequently be connected by tunnels.

One such model terminal for joint marine and rail use has been established through the enterprise of the Bush Company at South Brooklyn. It is the only example of an extensive and properly equipped terminal on the Atlantic seaboard in North America. Montreal is providing Canada with a similar terminal under the auspices of the Dominion Government. New Orleans and San Francisco are making creditable advances under City and State supervision, respectively, and in South America, Rio and Buenos Aires have been modernized. Aside from these cities, however, we must seek the northwestern ports of Europe for examples of properly equipped seaport terminals.*

---

* The following quotation from the report of the Boston Metropolitan Commission will ultimately be equally applicable to New York when freight tunnels under the river shall have become practicable:

> "Greater Boston being in effect one terminal district, the proper improvement and development of its terminal facilities should be treated as a single problem. The present arrangement of separate and independent freight terminals for the several railroads and lines of water transportation should be abandoned, and in its place there should be adopted a comprehensive plan for a flexible and efficient system of interchange terminals, whereby commodities, whether in carload or cargo bulk, or in lesser lots, may be delivered and transshipped from railroad to railroad or from land to water, or vice versa, and distributed among consignees, with the highest attainable expedition and economy. The establishment of such a system of interchange terminals is of the greatest importance to Boston. The controlling features of such a system should be direct and effective connection between all lines of railroad entering the city and every portion of the terminal system, as well as access to and use of the system and its facilities by every railroad on equal terms for like service."

South Brooklyn Terminal Site.

New York City has secured needed waterfront lands on both sides of the Bush Terminal at South Brooklyn, and purposes to add to its holdings, which should finally be improved by piers, warehouses and railroad equipment on the Bush model, and in such a manner as to supplement and not injure that enterprise. This can be done by the extension of the railroad back of the docks, from the Pennsylvania terminal at Bay Ridge, through Second avenue, extended to Gowanus Creek; across Gowanus, through the Erie Basin district, back of the New York Dock Company's properties to Atlantic Basin; provision to be made for a general assembly and transfer yard for less than carload freight in the vicinity of Erie Basin. Such an improvement will greatly stimulate the industrial development of a large section of South Brooklyn now fallen into comparative decay as a consequence of insufficient railroad connections.

If the United States Government shall not permit pier lines to be further extended in Manhattan, the largest class of ocean steamers will probably be obliged to dock in the future either in the South Brooklyn or Staten Island districts. Such location would afford comparatively cheap dockage and the dangers of navigating the upper bay would be avoided.

The State should locate one of the principal Erie Canal terminals immediately north of the City's property at the mouth of Gowanus, and make it an operative part of the general improvements contemplated.

Other Possible Terminal Sites.

Other strategic points for terminals about the harbor—to be connected by river tunnels in the future—are to be found in the vicinity of Stapleton and the Fresh Kill districts of Staten Island; at Newtown and Flushing Creeks; Jamaica Bay; inland along the line of the New York Connecting Railroad on Long Island; along the west side of Manhattan, and in the Morrisania and Port Morris districts, and should include a canal port of call at Spuyten Duyvil.

It is also desirable that terminal facilities, especially designed for the New England traffic, should be provided, if possible, with rail connections on the east side of Manhattan Island. This matter will be more fully discussed in my Annual Report.

Minor terminals and open public piers, to be used by car floats, scows, lighters, trucks, small craft and occasional ocean vessels, should be provided at frequent intervals about the commercial districts of the Port. Lighters are harbor trucks, and their use should be extended by accessibility to public docks and sheds. Small, inexpensive bulkheads and piers for the reception of coal, building materials and coarse freights generally, together with the raw materials for manufacture, should be built by the City where needed in outlying suburban districts, otherwise the cost of truckage will continue to remain an undue burden upon the community.

In New Jersey a comprehensive and more extensive system of terminals than that which now exists should be provided by the railroads for the assemblage and interchange of less than carload freight; for car transfer and car storage. This will greatly facilitate the use of small assembling lighterage terminals, both in New Jersey and in New York, to the advantage of both States. These matters will be discussed more at length in my Annual Report.

Conditions Along the North River Waterfront.

The demand for more and better pier and terminal facilities along the west side of Manhattan Island is steadily increasing; the Dock Department is constantly in receipt of requests for additional piers; the demand for open piers for the receipt of coal, building materials and coarse freight, and for the use of small steamboats, is insistent. Complaints of delays and difficulties due to congestion of trucks along the marginal way are also increasing. More piers cannot be provided, because the space is wholly occupied, and the only solution of the problem of increasing the commerce of Manhattan, or indeed of maintaining it, is to develop some policy whereby the waterfront can be used to greater advantage. Hereafter the waterfront should be treated as a whole and there should be some relation of use maintained between all the piers and between the piers and back lands, as far as this is possible.

Congestion along West street is due principally to the fact that there is no method of transferring freight between the piers and warehouses or factories except by trucks. Very serious delays are occasioned, since it is necessary to utilize the piers for temporary storage of materials; this, of course, interferes with free and advantageous use of the pier as an unloading or transfer station.

The wide marginal way necessary for trucking, and unprovided with rail service, disadvantageously cuts off the back land from the docks and prevents its proper commercial or industrial use. The separated parts should be so organized for joint use that warehouses and factories may supplant barber shops, lodging houses and saloons.

From the Battery to Canal street the piers are generally in the hands of railroad companies. From Canal street to Twenty-third street the piers are occupied by the trans-Atlantic and coastwise steamship companies. From Twenty-third street to Seventy-second street some steamship companies are located; but a large part of this space is occupied by the New York Central and Hudson River Railroad Company. Although the waterfront is to some extent divided into railroad and steamship zones; yet there is no definite terminal development, and the piers are generally used without reference to each other.

The policy of the City has been to execute long term leases to railroad and steamship companies, usually on favorable terms to the City, but there has been no definite, comprehensive terminal development. As a consequence of its unique natural advantages the commerce of the City has continually expanded in spite of the lack of municipal progress and foresight; but the time has come when these natural advantages must be organized and modernized if the City is to hold its own against improved organization at other cities.

One purpose of this report is to outline a plan whereby the railroads can be assembled within some definite zone, which shall be devoted exclusively to a joint railroad terminal installation to be participated in by the railroads generally, and to include piers connected by rail to yards to which trucks could come; to storage warehouses and factories located directly inshore of the piers, and thence by rail to transfer yards where materials from piers, warehouses or factories could be placed in cars loaded for destination, and these cars taken to adjacent car floats, thus providing for the transfer, storage and fabrication of raw materials, and at the same time utilizing the waterfront in a more effective manner. By entering into a plan of this kind, having mutual advantages, the railroads would be able to release to the City certain piers now occupied by them in lower Manhattan, which could then be used with great advantage for open wharfage or for steamship purposes, and so relieve the inconveniences of congestion where it is most acute.

The New York Central & Hudson River Railroad has a rail connection from Spuyten Duyvil along the Hudson River to its yard south of Seventy-second street, which yard occupies the waterfront from Seventy-second to Sixtieth streets.

It also maintains tracks on Eleventh avenue at grade from Sixtieth street to its yard lying between Thirty-seventh and Thirtieth streets, and south of Thirtieth street on Tenth avenue and West street at grade to its yard at St. John's Park. These tracks have long been a nuisance and menace to public safety, and there is no doubt that they will have to be removed from the street surfaces. Such removal and the continued operation by the New York Central & Hudson River Railroad Company of its present terminals is provided for in the installation of this joint railroad terminal.

Proposed Location.

It does not seem practicable to locate the main terminal south of Twenty-third street, as the property back of the piers will no doubt be required for office buildings and the marginal way is necessary for trucks.

North of Seventy-second street the land immediately back of the waterfront rises abruptly and is devoted mainly to park and residential purposes and is not available for terminal buildings such as will be required. The available site, therefore, appears to be between Twenty-third and Sixtieth streets; and as the industrial and trucking district now lies south of Twenty-third street, the site chosen is that lying as near Twenty-third street as possible. On account of the angle which the marginal way makes with the cross streets south of Thirtieth street the blocks between Twenty-third and Twenty-fifth streets are shortened to such an extent that they will be too small for the required terminal buildings, and the south limit of the terminal is therefore taken at Twenty-fifth street.

At the present time the Baltimore & Ohio Railroad occupies the block between Twenty-fifth and Twenty-sixth streets; the Lehigh Valley Railroad the block between Twenty-sixth and

Twenty-seventh streets; the Erie Railroad the block between Twenty-eighth and Twenty-ninth streets, and the Pennsylvania Railroad the block between Thirty-seventh and Thirty-eighth streets, all situated between Eleventh avenue and the marginal way. These spaces are used for switching yards and yards devoted to the interchange of freight between cars and trucks, and each of these railroads operates car float bridges at the bulkheads directly in front of its yard. Between Thirtieth and Thirty-seventh streets, west of Eleventh avenue, the New York Central and Hudson River Railroad and the West Shore Railroad have terminal yards with freight sheds, and enjoy the additional advantage of a rail connection with the yard lying between Sixtieth and Seventy-second streets, and thence north to their main line at Spuyten Duyvil.

Between Thirtieth and Thirty-second streets the New York Central yard extends back practically to Tenth avenue, and between Twenty-ninth and Thirtieth streets almost to Ninth avenue. It is understood that the New York Central & Hudson River Railroad Company has in view the installation of a terminal development for its own use very much along the general lines of the individual terminals described in this report.

Description of Proposed Installation.

While it is not possible at this time to absolutely define the installation which will be most effective, and while it is probable that the various railroad companies entering into this scheme will have different ideas as to the installation, yet it is desirable so far to outline such a plan, therefore, as may, at least, serve as a basis for discussion.

A unit of any proposed installation should consist of a large double deck pier with the necessary freight handling appliances, having railroad tracks on the lower deck, and propably on the upper deck; bulkhead shed for distribution and transfer of freight; a loading yard close by for the transfer of freight between trucks and cars; a transfer yard for interchange of freight between cars; a space for the storage of large quantities of freight and raw materials and ample factory accommodations for the fabrication of raw materials.

The plan suggested shows for each railroad, or for a combination of railroads, a double decked pier and shedded bulkhead, both connected by rail with a building directly in the rear of the bulkhead. This building would extend from the marginal way to Eleventh avenue, and a similar building is planned for the block between Eleventh and Tenth avenues. The ground floor and the second floor are to be used to handle the freight between cars and trucks; to transfer freight between cars, or to transfer freight by means of elevators to the floors above. The third, fourth and fifth floors are to be devoted to storage warehouses and the upper floors to factory purposes.

As a part of this proposed terminal development, it will be necessary to provide a storage yard for cars, and a yard for bringing freight in less than carload lots to be distributed and loaded for destination. Such a yard is planned between Thirty-eighth and Fortieth streets. Sufficient transfer bridges will be placed at the bulkhead directly in front of this yard, so that immediately as cars are loaded they can be sent out. The cheaper lands of New Jersey will be availed of, as now, for additional car storage.

It will be noted that the opportunity exists for expansion in this terminal development to the north of the yard between Thirty-eighth and Fortieth streets, as far, say, as the park at Fifty-fourth street.

A four track elevated railway is planned from Twenty-fifth street along the marginal way to the storage yard between Thirty-eighth and Fortieth streets, and connections will be made from this elevated railway to each of the terminal buildings as well as to the existing yard of the New York Central and Hudson River Railroad north of Thirtieth street. This elevated railroad can be continued north along the marginal way to connect with the yard at Sixtieth street, and with this in operation there will be no further necessity for surface tracks on Eleventh avenue.

In connection with this proposed terminal development, an extension of the elevated railway south of Twenty-fifth street, having spurs to many of the piers, would seem desirable.

It is also desirable that warehouses should be located on the easterly side of the marginal way below Twenty-third street for the receipt of cargo coming in by steamships. Connections can be made from the elevated railway to the second floor of these warehouses, so that a convenient interchange of freight can be made between the proposed steamship terminals and the joint railroad terminal.

The New York Central and Hudson River Railroad Company

maintains a surface railway along the marginal way to its terminal yard at St. John's Park, south of Canal street. If an elevated railway should be substituted for this it will be difficult and expensive to maintain the connection with the St. John's Park terminal as a consequence of passage over the Ninth Avenue Elevated Railway. To furnish facilities for the downtown districts a joint terminal yard is planned north of Canal street. Property will be acquired from the marginal way back to Washington street, between Spring and West Houston streets, and an elevated railway yard will be built, provided with platforms and elevators to transfer the freight to and from the surface, where it can be transferred to and from trucks. The present surface railway will then be discontinued.

The elevated railroad plan is solely for the transportation of freight, and should not be complicated by attempts to provide for passenger traffic.

It will be noted on Plate I that provision is made at Twenty-fourth street and at Forty-first street for properly equipped piers for open wharfage and canal boats.

Piers.—The piers will be 60 feet wide, with a slip space of 150 feet between piers; their length will be approximately 600 feet and they will be double decked. The lower deck will have a direct rail connection by two tracks along the centre, or side lines of the pier, as is found most advantageous, with the connecting tracks along the marginal way and with those entering the buildings. The second deck will probably have similar connections with the elevated railway and the second floors of the buildings. All materials will be loaded by hand trucks and mechanical apparatus on box cars or flat cars provided with removable canvas tops. These cars will be taken into the first or second stories of the building in the rear, or to the back streets, where trucks can receive the materials directly from the cars. No horse trucking whatever will be permitted west of the easterly side of the marginal way.

A recreation structure will be provided above the second deck of each of the piers, which can be reached from Tenth avenue or Eleventh avenue by elevated passageways and overhead bridges. This will serve the needs of the great industrial population which will be attracted to the vicinity, while preventing the users of the recreation structures from coming in contact with the railroad tracks or the operating floors of the piers. The piers in front of the terminal buildings should be distinctly under the control of the respective railroads and may be used for steamships, lighters or in any other way the railroad company finds advisable; the general control, however, to be subject to the approval of the terminal corporation, referred to elsewhere.

Bulkheads.—A bulkhead shed with a platform 30 feet wide will be constructed along the entire length of the pier installation, and freight can be transferred directly from this bulkhead to flat cars alongside.

Marginal Way.—The marginal way between the bulkhead sheds and the terminal buildings will be utilized by surface and elevated tracks exclusively, to the exclusion of trucks.

Terminal Buildings.—The difference in grade between Tenth avenue and the marginal way is such that the tracks from the elevated railway which enter the second story of the terminal building will cross over Eleventh avenue, and descending on a slight grade, will reach Tenth avenue at about the grade of the street, so that trucks may enter the building at Tenth avenue on what is really the second floor.

The first floor of the building between Tenth and Eleventh avenues and between Eleventh avenue and the marginal way will be entered by trucks at the grade of Eleventh avenue, and the floor will be depressed, so that it becomes a basement floor at Tenth avenue.

The building between Tenth and Eleventh avenues will be 800 feet long and 200 feet wide, and between Eleventh avenue and the marginal way will be 200 feet in width and in length will vary from 800 feet to about 550 feet, due to the angle in Twelfth avenue.

A track immediately adjoining the building with a platform alongside is planned, so that, in addition to the space within the buildings, platforms shall be provided for the transfer of freight between the flat cars and trucks outside the buildings.

On the roof over this platform, which is outside of the building, a footwalk is provided leading from Tenth avenue to the marginal way, and thence by elevated footway across the elevated railway to the recreation structures on the piers.

The building as planned will be of fireproof construction and the insurance rates should be low.

A power house will be constructed which will provide light, heat and power for the building as well as for the operation of electrical traction on the railway. It will, of course, be impracticable to use steam locomotives inside of the buildings.

Plate 2, Figure 1, shows a detail of the ground floor of these buildings, and as indicated, this floor is reserved for railroad tracks and trucks, and for the handling of freight between the cars and trucks. The elevators which are shown on the platforms are for handling freight between the cars or trucks and the warehouses or factories above.

The second floor, as shown on Plate 2, Figure 2, is essentially similar to the first floor. On the southerly inside of each of these buildings a ramp is provided so that the cars may be transferred from the upper level to the lower level.

Plate 2, Figure 3, shows the storage floors, which may de divided in any manner considered desirable.

Plate 2, Figure 4, shows the factory floors. It will be noted that light wells are provided so that these floors will be very well adapted to factory purposes.

The very great advantage of having adjacent to the piers where the raw materials are to be received a place to store them until such time as they shall be required, and the very great convenience of the factory immediately overhead can be easily appreciated. This will do away with a very large amount of trucking and street congestion, which is expensive and the cause of very serious delays.

Storage Transfer Yard.—The storage yard between Thirty-eighth and Fortieth streets is shown on Plate 3. It has three levels, with ramps for the transfer of cars between the different levels. The lower, or street level, as shown on Plate 3, Figure 1, is provided with transfer sheds which will be used for assembling freight received in less than carload lots to be loaded into cars for destination. The float bridges are shown directly in front of this yard.

The second, or elevated railway level, as shown on Plate 3, Figure 2, of the yard, is practically the same plan as provided for the first floor. It is contemplated that mixed freight will be brought to this level from the terminal building and will be transferred to the lower floors by chutes, or elevators, and there loaded for destination. In the afternoon, when a large volume of business will be done at this yard, the second floor will be occupied with cars coming from the terminal buildings, and on the lower level the cars will be loaded as rapidly as possible and sent out to the car floats, avoiding interference between cars entering and leaving the yard.

The third level of this yard, as shown on Plate 3, Figure 3, will be used entirely for the storage of cars. Materials which come from the terminal buildings or the piers, loaded in carload lots, will not necessarily be taken into the yard, but may be dispatched directly to the car floats.

Plate 4 shows ground level plan and elevated floor plan of the proposed transfer station between Spring and West Houston streets, equipped with assorting platforms and elevators for handling freight between the two levels.

Plate 5 shows the present occupancy of the blocks from West Twenty-third street to West Seventy-second street, and between the marginal way and Eleventh avenue.

Provision is made in this installation for handling the following:

General merchandise delivered to the piers from vessels to be taken away on trucks.

General merchandise delivered to the piers to be reshipped by the railroads.

General merchandise delivered to the piers to go into storage.

General merchandise delivered to the piers for factory purposes.

General merchandise delivered to the piers for reshipment on ships.

General merchandise delivered on trucks for export on the piers.

General merchandise delivered on trucks for railroad shipment.

General merchandise delivered on trucks for storage or factory purposes.

General merchandise delivered on cars for export on ships.

General merchandise delivered on cars for re-railroad shipment.

General merchandise delivered on cars for trucks.

General merchandise delivered on cars for storage or for factory purposes.

General merchandise in storage or factory for railroad shipment.

General merchandise in storage or factory for trucks.

General merchandise in storage for factory purposes.

General merchandise in factory for storage.

General merchandise in storage or factory for export from the piers.

The elevated railway connection between Fifty-fourth and Sixtieth streets might properly be constructed and operated by the New York Central and Hudson River Railroad Company. Provision for additional tracks on the Riverside Drive section should receive especial consideration under a plan which shall provide that the railroad will cover the tracks, the deck to constitute the new surface of the park, thus eliminating the tracks from view. The New York Central being the only western line having direct rail access to Manhattan, should be permitted every advantage which its favorable position allows of, always provided the interests of the City are thereby subserved.

The advantages of the water level route of the New York Central lines, under the stimulus of the Erie Canal improvements, should provide the basic railroad rate for the United States. Rates from Chicago north to Montreal and south to New Orleans will be influenced by this competition, as well as rates eastward from the Mississippi to all Atlantic seaboard cities. Manhattan will be placed at a serious disadvantage with New Jersey and other parts of the Port if the New York Central's direct rail connection with the West shall be interrupted or shall not be taken full advantage of. In fact the surface tracks cannot be eliminated until other tracks available for other roads leading to the West shall be provided. When this policy shall have been established by the City, then the Central must perforce submit to it, since it can have no monopoly of occupancy. Let the City's intention be determined and compliance with it by the New York Central will be prompt. The City will not be obliged to wait for the physical completion of the improvements before the Central will substitute overhead for surface transit.

The City should itself work out its local transportation problems, without depending too much upon suggestions and advice from the railroad companies, each naturally looking after its own separate interest.

Experience has demonstrated that the railroads cannot provide adequate City terminals without municipal co-operation:

1st. Their rivalries are such that, unless some outside unifying and organizing influence is brought to bear, the roads will not act in concert.

2d. It is essential to have such concerted action, both for economies and for purposes of joint occupancy, under which no invidious discrimination between roads shall be possible.

3d. The condemnation of separate land units for operation will entail great expense as compared with organized condemnation for joint use, and may compel the railroads to utilize the land so taken for transportation purposes only, thus losing the advantage of rentals of overhead floors for industrial purposes, which rentals are essential to make the enterprise financially self-sustaining.

4th. It is very doubtful if the public opinion of this City will permit the utilization of the marginal way exclusively for track purposes, as here contemplated, under private as distinguished from public railroad occupancy.

Financial Plan.

As the result of recent legislation releasing self-sustaining dock bonds from inclusion in the debt limit, I believe it possible that terminal facilities can be financed on any scale applied for by the railroads without an excessive burden being imposed either upon the City or upon them. If assurance can be had that the enterprise will be self-sustaining and capable of amortization within a reasonable time, then consideration of cost is comparatively immaterial, since municipal bond issues would be promptly exempted from the debt limit as soon as made, and the resources of the City available for other kinds of improvements would not be adversely affected.

The installation may be undertaken either on a grand scale, with provision made for all carriers, and for storage and industrial uses, which shall assume community proportions and involve social and economic changes of great moment; or again, in case of non-participation by the carriers generally, or of civic timidity to act in a large way, it can be shrunk to provisions for terminals for a few roads, or even to the limitation of a west side municipal elevated railroad for joint use, with provision for private terminals by participating railroads. If necessary the New

York Central alone could be provided for in this manner through the instrumentality of a contract based on a reasonable service charge. It is reasonable to assume, however, that no railroad will neglect an opportunity to obtain terminal and transit facilities on the west side of Manhattan on a basis advantageous to itself. The larger the terminal and its adjuncts the better will it serve the City's needs and the less burdensome will it be to the participating roads. In any event the construction of a municipal elevated freight road should not be delayed.

The plan as here outlined contemplates initial municipal expenditure for lands, docks, elevated railroad structure, upper and lower terminals sufficient for all roads, except the Central, which company already owns lands above Twenty-third street which it can utilize jointly or separately for terminal purposes. Should the undertaking assume proportions beyond those of a mere elevated railroad, and should provision be made for ample overhead space for general storage, factory and warehouse purposes, rentals will go far to meet expenses, since no similar conveniences can be obtained elsewhere in the Port. At the end of the contractual amortization period the terminal will become the City's property. Not only will the City be working toward such exclusive ownership, but in the interim terminal facilities which are necessary to maintain the supremacy of the Port and to overcome the insular disadvantages of Manhattan will have been provided.

As a necessary preliminary to such an undertaking, the City, after negotiations with the railroads, should seek to procure the incorporation of a terminal company in which the roads shall participate as stockholders, and on the board of trustees of which the City shall be represented by its three general officers and the dock commissioner—for the purpose of securing publicity and of influencing operation and future policy.

Each railroad company will pay to the terminal company a proportionate rental to be agreed upon; the total amount thus paid to the terminal company must be guaranteed in advance by the participating roads to yield 5 per cent. annually to the City on its investment, after charges for administration, maintenance and operation shall have been met. For the moneys which the railroads pay they will receive from the terminal company the use of the piers, terminal, tracks, float bridges, and the two lower floors of the buildings for track purposes, together with light, heat and power. They themselves will individually arrange for the occupancy of the upper floors of the buildings, and at uniform rates receive rentals therefor. All sub-rentals of floor space above a fixed amount—which shall be determined upon as an offset to counteract the charges of the City for the use of the property—shall be paid by the roads to the terminal company, and by it to the City. This will form a fund for the extension of terminal improvements at such times and at such points as the City may determine.

The City would then proceed to acquire the necessary property, construct and equip the buildings, docks and elevated railroad with the money received from its bonds, which would be marketed presumably on a 4 per cent. basis; the difference of 1 per cent. being sufficient to amortize the cost in forty-one years, the lease of the property to the terminal corporation to remain in continuance for such period. The terminal company would then administer the property.

The large amount of New York Central property above Twenty-third street may be included in the above joint terminal plan, if the company should so elect. If not, the enterprise could be proceeded with without being participated in directly by the Central except that its use of the connecting elevated railroad would only be permitted on such terms as should be fair and reasonable to the other participating roads, and as would promote the general harmonious and efficient use and occupancy of the joint terminal as a whole.

The car capacity within the terminal buildings would be 2,800 cars; in the storage and distributing yard between Thirty-eighth and Fortieth streets, 1,850 cars; in the downtown yard, 330 cars.

It is probable that it will be necessary to increase the storage capacity of the downtown yard unless additional elevated railway car storage can be provided by a wider elevated structure than that planned over the marginal way.

The total floor space for storage of commodities is approximately 3,600,000 square feet, and the total space available for industrial use is 7,400,000 square feet, or a total of 11,000,000 square feet.

The enterprise if carried out as planned in its entirety would cost approximately $100,000,000.

This figure is the result of a general study of the situation, and includes both the cost of construction and the acquisition of property. It is given to serve as a basis for public discussion and negotiation with the railroads, upon whose final judgment and co-operation the undertaking must depend for success.

Summary.

The common interests of the carriers and the City for modern terminals is so great—the complex difficulties of obtaining them has so far outgrown the ability of each to act separately—that selfish individual corporate action, on the one hand, and popular prejudice against corporations, on the other, must give place to a mutually helpful promotion policy which shall seek to utilize all possible resources.

The immense capital invested in ocean steamships of the largest size is such as to make imperative terminal facilities located and equipped for the utmost possible dispatch for loading and unloading cargo, otherwise economies in transit will be absorbed in needless port delays. The State and United States governments should also co-operate to this end by affording every facility which the customs and quarantine services have at their disposal to promote dispatch.

Rail freight rates for transit are not usually segregated from terminal charges. The freight rate includes both services. Consequently, as between cities having identical freight rates, those whose terminal service is quickest and cheapest will attract commerce away from others.

As noted at the beginning, the above report is submitted only as a basis for criticism and for negotiation with railroad and steamship companies seeking better terminals. I have sought to present the larger general features of the problem, which, in brief, consist of properly articulating warehouses and factories, steamships and railways. Other ports, and the New Jersey district of this port, will gain what New York will lose if it shall fail to do this.

The one cardinal principle to keep in mind is the quick and cheap exchange of commodities between water and land vehicles, warehouses and factories, eliminating the expensive use of the truck, with its attendant street congestion.

Finally, public opinion must be aroused to the necessity of

modernizing the Port, and a comprehension of how this can be accomplished must be made clear by discussion.

Respectfully submitted,

CALVIN TOMKINS,
Commissioner of Docks.

NOTE.

Provision for joint terminals adapted for industrial use on the west side of Manhattan will tend to increase street and tenement congestion, and this should be taken into serious consideration now.

Present conditions need not be made worse; on the contrary, systematic use of the blocks west of Ninth avenue might accommodate the terminals proposed with the manufacturing and residence development thus attracted without the danger or discomfort of present conditions; but to accomplish this, intelligent planning must be had.

Under present conditions, and as a consequence of increasing cost of deliveries, manufacturing industries are migrating from Manhattan to Long Island, New Jersey and elsewhere, where railroad sidings are available and cheap delivery charges attainable.

This exodus of capital and population will be checked by the terminal industrial reorganization of Manhattan, which borough will then of necessity be obliged to more rapidly reorganize its street and park systems by widenings, extensions and provisions for small parks. These changes can only be accomplished by making this class of improvements pay for themselves through better condemnation laws and more frequent resort to the processes of beneficial assessment and of excess condemnation. A constitutional amendment to make this last procedure available has passed the Legislature this year, and its passage next year and subsequent availability through enabling legislation should be encouraged by the City.

Store door deliveries of higher class freight by railroad and steamship vehicles will tend to relieve congestion in the streets and at the terminals, by substituting an organized delivery service for the present individual unrelated service. It will also soon become necessary to regulate traffic in the harbor waters as street traffic is now regulated, to avoid danger, expense and delay.

Provision should also be made for securing adequate light and circulation of air in the streets, which can be accomplished through the establishment of an angle of light in the streets, as is the practice in European cities, without unduly limiting the height of tall buildings. It is immaterial how tall a building is, provided the street lights are preserved. Interference of buildings with each other, as regards light and air, on other than street lines, can safely be left to private interest. If the street lights are preserved the tendency will be more and more to utilize the *city block* instead of the *city lot* as the unit for tall building construction.

There should be a progressive increase of the incidence of taxation upon land as distinguished from buildings and other improvements on land.

Subways must of necessity be rapidly developed to disperse excess population to suburban districts. To accomplish this the City must retain control over new subways at the center—municipalizing them instead of permitting them to pass under private control. Suburban extensions should be paid for by local assessment upon property benefited, so that such extensions may be *promptly* had without involving unduly high rates of fare. If the City shall now establish this policy of complete municipalization and make it alternately possible to operate new subways as extensions of existing concessions—or independently—the present operating corporations must perforce take them over on a profitable basis for operation, which the City should seek to provide. Complete City monopoly in extensions and partial monopoly in service—also gradually to be made complete—should be the end sought.

The reconstruction of the old sewer system of Manhattan should be considered in relation to subway and waterfront improvements, otherwise great and costly confusion will arise. East and West Manhattan subways should be superimposed on North and South lines. A comprehensive subway plan for the entire City should be provided without further delay. A Municipal subway *plan* and *policy* are as necessary at this time as a municipal plan and policy for the handling of freight.

The greatest problem—already showing itself on every hand—is that of so co-ordinating all public services as to make each best serve the other and all best serve the City. Here again progressive monopoly in City control is the essential. Whether this shall be economically developed through intelligent co-operation of present public service corporations—or the City's hand forced by their obstructive tactics—is a question of first importance.

www.ingramcontent.com/pod-product-compliance
Lightning Source LLC
LaVergne TN
LVHW020632110826
845149LV00004B/1149